Operations

Algebra

Mathematics in Context

ENCYCLOPÆDIA Britannica

Mathematics in Context is a comprehensive curriculum for the middle grades. It was developed in 1991 through 1997 in collaboration with the Wisconsin Center for Education Research, School of Education, University of Wisconsin-Madison and the Freudenthal Institute at the University of Utrecht, The Netherlands, with the support of the National Science Foundation Grant No. 9054928.

The revision of the curriculum was carried out in 2003 through 2005, with the support of the National Science Foundation Grant No. ESI 0137414.

National Science Foundation
Opinions expressed are those of the authors and not necessarily those of the Foundation.

© 2010 Encyclopædia Britannica, Inc. Britannica, Encyclopædia Britannica, the thistle logo, *Mathematics in Context*, and the *Mathematics in Context* logo are registered trademarks of Encyclopædia Britannica, Inc.

All rights reserved.

No part of this work may be reproduced or utilized in any form or by any means, electronic or mechanical, including photocopying, recording or by any information storage or retrieval system, without permission in writing from the publisher.

International Standard Book Number 978-1-59339-940-5

Printed in the United States of America

2 3 4 5 C 13

The *Mathematics in Context* Development Team

Development 1991–1997

The initial version of *Operations* was developed by Mieke Abels and Monica Wijers. It was adapted for use in American schools by Gail Burrill, Aaron N. Simon, and Beth R. Cole.

Wisconsin Center for Education Research Staff

Thomas A. Romberg
Director

Joan Daniels Pedro
Assistant to the Director

Gail Burrill
Coordinator

Margaret R. Meyer
Coordinator

Freudenthal Institute Staff

Jan de Lange
Director

Els Feijs
Coordinator

Martin van Reeuwijk
Coordinator

Project Staff

Jonathan Brendefur
Laura Brinker
James Browne
Jack Burrill
Rose Byrd
Peter Christiansen
Barbara Clarke
Doug Clarke
Beth R. Cole
Fae Dremock
Mary Ann Fix

Sherian Foster
James A. Middleton
Jasmina Milinkovic
Margaret A. Pligge
Mary C. Shafer
Julia A. Shew
Aaron N. Simon
Marvin Smith
Stephanie Z. Smith
Mary S. Spence

Mieke Abels
Nina Boswinkel
Frans van Galen
Koeno Gravemeijer
Marja van den Heuvel-Panhuizen
Jan Auke de Jong
Vincent Jonker
Ronald Keijzer
Martin Kindt

Jansie Niehaus
Nanda Querelle
Anton Roodhardt
Leen Streefland
Adri Treffers
Monica Wijers
Astrid de Wild

Revision 2003–2005

The revised version of *Operations* was developed by Martin Kindt and Truus Dekker. It was adapted for use in American schools by Gail Burrill.

Wisconsin Center for Education Research Staff

Thomas A. Romberg
Director

David C. Webb
Coordinator

Gail Burrill
Editorial Coordinator

Margaret A. Pligge
Editorial Coordinator

Freudenthal Institute Staff

Jan de Lange
Director

Truus Dekker
Coordinator

Mieke Abels
Content Coordinator

Monica Wijers
Content Coordinator

Project Staff

Sarah Ailts
Beth R. Cole
Erin Hazlett
Teri Hedges
Karen Hoiberg
Carrie Johnson
Jean Krusi
Elaine McGrath

Margaret R. Meyer
Anne Park
Bryna Rappaport
Kathleen A. Steele
Ana C. Stephens
Candace Ulmer
Jill Vettrus

Arthur Bakker
Peter Boon
Els Feijs
Dédé de Haan
Martin Kindt

Nathalie Kuijpers
Huub Nilwik
Sonia Palha
Nanda Querelle
Martin van Reeuwijk

Cover photo credits: (left to right) © Getty Images;
© John McAnulty/Corbis; © Corbis

Illustrations
1, 2 Holly Cooper-Olds; **6–8, 12, 16–18, 31, 32, 44, 46** Christine McCabe/
© Encyclopædia Britannica, Inc.

Photographs
2 © PhotoDisc/Getty Images; **8** (left to right) Photo by Bill Middlebrook - Breckenridge, Colorado; © David Muench/Corbis; **9** © PhotoDisc/Getty Images; **13** (top) © Corbis; (bottom) © PhotoDisc/Getty Images; **22** Brand X Pictures/Alamy; **28** © EB Inc.; **30** (left) © Corbis; (middle and right) © PhotoDisc/Getty Images; **32, 33** © PhotoDisc/Getty Images; **36** Don Couch/HRW Photo; **37** © PhotoDisc/ Getty Images; **45** Joseph Sharp, Graphic Designer for Provo City

Contents

Letter to the Student   vi

Section A — Positive and Negative

What Time Is It There?   1
World Time Zones   3
Below and Above Sea Level   6
Summary   10
Check Your Work   10

Section B — Walking Along the Number Line

Ordering the Numbers   12
Ronnie the Robot   16
The Robot Game   18
Summary   20
Check Your Work   20

Section C — Calculating with Positive and Negative Numbers

Adding and Subtracting   22
The Integer Game   27
Temperatures and Altitudes   28
Higher and Higher   30
Summary   34
Check Your Work   34

Section D — Adding and Multiplying

Calculations Using Differences   36
Multiplication with Positive
  and Negative Numbers   38
Summary   42
Check Your Work   43

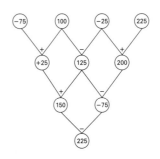

Section E — Operations and Coordinates

Directions   44
Changing Shapes   46
Summary   50
Check Your Work   51

Additional Practice   52

Answers to Check Your Work   58

Dear Student,

Sometimes it is necessary to have numbers that show different directions—or opposites.

Have you ever used positive and negative numbers?

In this unit, you will use a world map to explore time zones and figure out the best times to call people in other parts of the world. You will practice adding, subtracting, and multiplying positive and negative numbers in different contexts. Ronnie the Robot will help you to work with a number line. You will multiply and divide positive and negative numbers to find average temperatures.

In the last section, you will investigate how to move, enlarge, and reduce a shape on graph paper using positive and negative numbers.

We hope you enjoy this unit and learn a lot about operations with positive and negative numbers.

Sincerely,

The Mathematics in Context Development Team

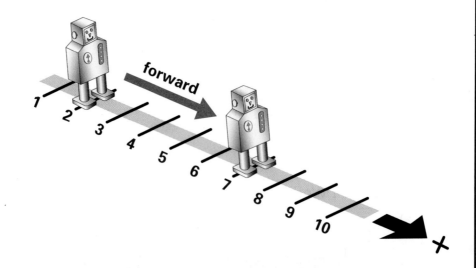

A Positive and Negative

What Time Is It There?

A. Harold and Felicia are big tennis fans. They are watching the Australian Open final being transmitted directly from Australia. Suddenly, their little brother enters the room.

B. It is 7:30 P.M., and Peter knows that tomorrow his cousin Susan is giving her first solo piano recital. He calls her in London to wish her luck.

A Fair

C. Mary is flying from Minneapolis to Seattle. Here is her flight information.

Flight	Date	From/To	Time
NW1607	12/05	Minneapolis to Seattle	11:30 A.M. to 1:00 P.M.

Mary is happy that her trip only takes $1\frac{1}{2}$ hours.

Flight	Date	From/To	Time
NW0008	12/12	Seattle to Minneapolis	11:45 A.M. to 5:15 P.M.

Mary wonders why the trip back from Seattle takes so much longer than the trip to Seattle.

1. These three stories have something in common. Can you explain what it is? This photo of the Earth might help.

2. **Reflect** Have you ever had an experience like those in the three stories? If so, describe it.

World Time Zones

Student Activity Sheet 1 shows a map that can be folded into a cylinder and a special ring called a **sun ring**. Use **Student Activity Sheet 1** to make the cylinder and the ring. The cylinder is a three-dimensional model of the Earth that can help you answer problem 3.

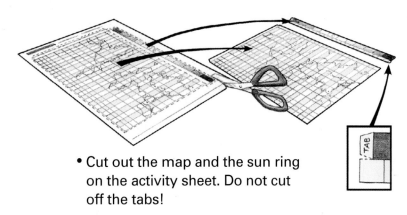

- Cut out the map and the sun ring on the activity sheet. Do not cut off the tabs!

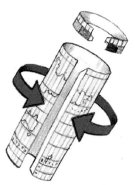

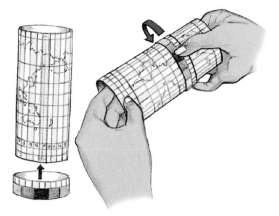

- Roll the map into a cylinder and tape it closed over the tab. The sections marked +12 and −12 should overlap. Do the same thing for the sun ring.

- Slide the sun ring over the cylinder.

3. **a.** How can the sun ring on your model explain the photo on the previous page?

   b. How can you use the sun ring on your model to explain the first and second stories on page 1?

Section A: Positive and Negative 3

A Positive and Negative

4. a. When it is noon in New York, where is it midnight?

   b. What is the time difference between New York and Los Angeles?

   c. What is the time difference between London and New York?

   d. What does it mean that Greenwich, England, is in the section marked with the number 0?

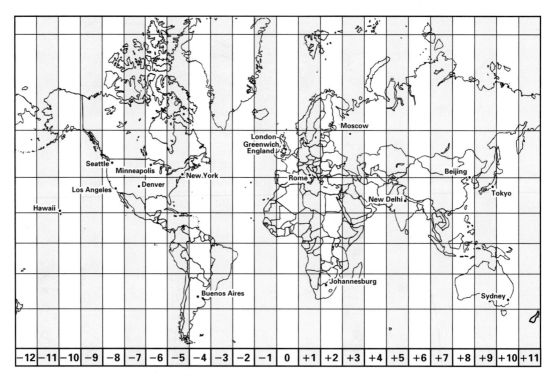

Note: This map is a simplified version of an actual time zone map, on which the zones often vary to accommodate islands, country borders, and certain geographical features.

This map shows the international **time zones**.

Time is calculated from a zero line in Greenwich. There are 24 zones. When you move from one time zone to the next, you have to change your watch by one hour either backward or ahead, depending on which way you are traveling.

A strip at the bottom of the map shows how the time zones are related to the zero-time zone. For instance, if the time in Greenwich is 9:15 A.M., it is already 10:15 A.M. in Rome (zone marked +1).

You may want to look at an atlas to see the actual time zones.

Positive and Negative

The continental United States has four time zones: Eastern, Central, Mountain, and Pacific.

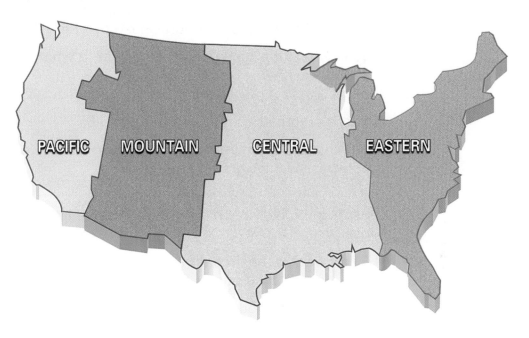

5. a. Compare the map above to your cylinder map from **Student Activity Sheet 1**. What numbers on the cylinder map correspond to the four time zones shown above?

  b. If you travel from one time zone east to the next time zone on the cylinder map, what happens to the numbers?

  c. If it is 11:30 A.M. Eastern time, what time is it in the Pacific time zone?

  d. When you travel east, should you change the time on your watch one hour backward or ahead?

6. a. On the cylinder map, find the time zone in which you live. What is the number for your time zone? What does this number tell you?

  b. In what time zone is Hawaii? What about Moscow?

  c. Name a city in the time zone marked +2 (which is read as "positive two"). What is the time difference between your time zone and the one in that city? Explain how you found your answer.

Section A: Positive and Negative 5

Positive and Negative

Tara, Victor, and José are classmates. The three students live in a city in time zone −5 (which is read as "negative five"). They have relatives who live in countries all over the world. Tara's cousin Keisha lives in a place where they are six hours ahead of where Tara lives.

7. In what time zone does Keisha live? In what countries might Keisha live?

Victor's grandfather lives in time zone +5.

8. What is the time difference between where Keisha lives and where Victor's grandfather lives?

José's uncle lives in a place where it is two hours earlier than where José lives.

9. a. What is the time difference between where Keisha lives and where José's uncle lives?

b. How did you find this difference?

10. At 4:00 P.M. Keisha wants to phone Tara. Is it a good time to call? Why or why not?

Numbers like +2 are called **positive numbers**.
Numbers like −5 are called **negative numbers**.

Below and Above Sea Level

Fleur's class in Nieuwerkerk, The Netherlands, regularly exchanges e-mails with Diego's class in Eagle, Colorado. Diego wrote this e-mail to Fleur:

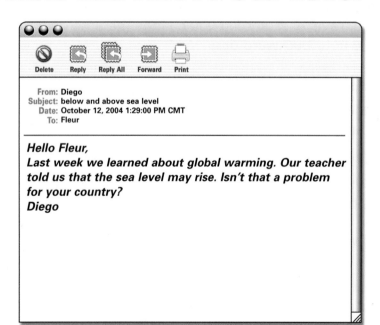

From: Diego
Subject: below and above sea level
Date: October 12, 2004 1:29:00 PM CMT
To: Fleur

*Hello Fleur,
Last week we learned about global warming. Our teacher told us that the sea level may rise. Isn't that a problem for your country?
Diego*

Positive and Negative

And Fleur answered:

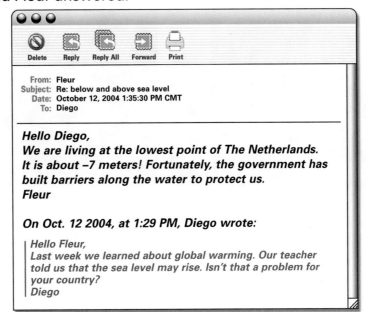

From: Fleur
Subject: Re: below and above sea level
Date: October 12, 2004 1:35:30 PM CMT
To: Diego

Hello Diego,
We are living at the lowest point of The Netherlands. It is about −7 meters! Fortunately, the government has built barriers along the water to protect us.
Fleur

On Oct. 12 2004, at 1:29 PM, Diego wrote:

Hello Fleur,
Last week we learned about global warming. Our teacher told us that the sea level may rise. Isn't that a problem for your country?
Diego

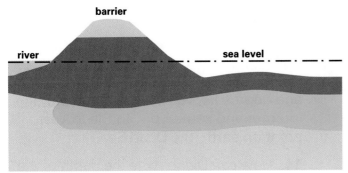

11. What does "−7 meters (m)" mean?

12. The height of the barrier Fleur mentioned is about 6 m above sea level. Find a short notation for "6 m above sea level."

In her next e-mail, Fleur wrote:

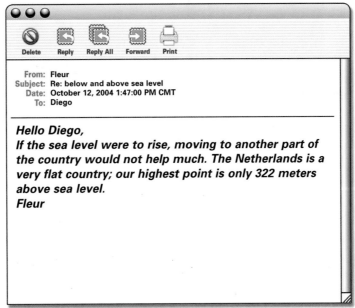

From: Fleur
Subject: Re: below and above sea level
Date: October 12, 2004 1:47:00 PM CMT
To: Diego

Hello Diego,
If the sea level were to rise, moving to another part of the country would not help much. The Netherlands is a very flat country; our highest point is only 322 meters above sea level.
Fleur

Section A: Positive and Negative 7

Positive and Negative

Diego answered that the highest point in Colorado, Mount Albert, has a height of about 4,400 m above sea level, and the lowest point in Colorado, on the Arkansas River, is still 1,021 m above sea level.

13. a. What is the difference in height between the highest point in Colorado and the highest point in The Netherlands?

   b. What is the difference between the lowest point in Colorado and the lowest point in The Netherlands?

Now Diego became interested in heights and depths as well. He searched the Internet and came up with the lowest point on earth, the Dead Sea.

14. How can Diego write the depth of the Dead Sea in a shortened way?

Positive and Negative

He also found the lowest point in the United States on the Internet. It is in Death Valley, California, and the depth is −282 feet (ft). Diego found that 1 ft = 0.3048 m.

15. Estimate the depth of Death Valley in whole meters. Use a correct notation.

Hint: A meter is about three feet.

Everybody is now interested in record highs and lows. The table below shows a list of heights and depths students found.

Name	Below or Above Sea Level
Florida	lowest: 0 m
Louisiana	lowest: −2.4 m
Alabama	highest: +733 m
Colorado	highest: +4,400 m
Washington, D.C.	lowest: +0.3 m
Nepal	highest: +8,850 m
Challenger Deep	lowest: −11,000 m
Europe	lowest: −28 m

16. What is the lowest point of Washington, D.C., expressed in feet?

17. a. Is the lowest point in Florida below, above, or at sea level?

   b. Why does the lowest point in Florida not have a plus or a minus sign?

Note: The Challenger Deep is the lowest point in the oceans of the earth. It is situated in the Pacific Ocean, near the Marianas Islands.

If you cut Mount Everest off at sea level and put it on the ocean bottom in the Challenger Deep, there would still be about a mile of water over the top of it!

Section A: Positive and Negative 9

Positive and Negative

Summary

You can use positive and negative numbers in many situations. In this section, you used them for time zones east from the zero line (+) and west of the zero line (−). You also used positive and negative numbers for above sea level (+) and below sea level (−).

Positive numbers are often written with a + in front of the number, but sometimes they are written without it. Either way, they mean the same thing. However, you must write a negative sign for a negative number.

Zero (0) is neither positive nor negative.

Check Your Work

1. Read the story about Mary's trip to Seattle on page 2 again. Did the trip back from Seattle really take longer? Explain your answer.

2. Write down two situations in which you could use positive and negative numbers. Explain how you would use them in the situations you described.

Use the list of heights and depths students found on page 9 to answer the following questions.

"Let's make a scale drawing showing all of the heights and depths from the lowest to the highest point," Erica suggests.

3. a. What is the distance (in meters) between the highest and lowest points on the scale?

   b. Could you make a scale drawing showing the highest and lowest points using a scale of 1:100? Why or why not? Remember that a scale 1:100 means that 1 centimeter (cm) in the drawing equals 100 cm (or 1 meter) in the actual situation.

10 Operations

Jassir used positive and negative numbers to show how many meters a trail in the mountains goes up and down. Here is Jassir's table for the Mirror Lake Trail.

Mirror Lake Trail Uphill/Downhill (in m)
+230
−130
+37
−340
+110
−37
+140
−40

4. a. Estimate whether you would end up higher or lower than where you started if you hiked this trail.

Jassir uses the following method to find out exactly how many meters higher or lower the endpoint is.

> You can cancel +37 uphill and −37 downhill. Some other numbers can be combined too.

b. Find out how much lower the end point is compared to the starting point for Mirror Lake Trail. You may use Jassir's method.

For Further Reflection

Explain why labeling some numbers positive and some negative is helpful. Think of some situations different from those in this section in which you would use positive and negative numbers.

Section A: Positive and Negative 11

B Walking Along the Number Line

Ordering the Numbers

At the bottom of the time zone map, you saw a strip with 24 numbers, from −12 to +11.

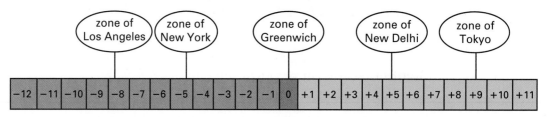

1. What is the time difference between New Delhi and New York? Between New York and Los Angeles?

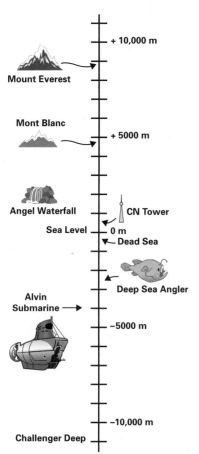

Number strips or **number lines** can be used for other purposes. For instance, this number line shows heights and depths.

2. a. Mount Everest in Nepal is the highest mountain on earth. From the table on page 9, you can read that its height is 8,850 m. About how many feet is that?

b. The highest mountain in Western Europe is Mont Blanc in Switzerland. About how many meters high is it?

c. At about what depth does the Deep Sea Angler live?

3. What is the difference in height between Mount Everest and Challenger Deep?

12 Operations

Walking Along the Number Line

4. a. Read the following three statements. Do you agree with them? Explain why or why not.

- In the time zone for New York (−5), it's always earlier than in the zone for Moscow (+3).

- The lowest point in Louisiana (−2.4 m) is lower than the lowest point in Washington, D.C. (+0.3 m).

- If the high temperature was −10 degrees Celsius (−10°C) on Sunday and −2°C on Thursday, it was colder on Sunday than on Thursday.

b. Write three true statements like those above: one comparing times, one comparing altitudes, and one comparing temperatures.

Walking Along the Number Line

The statements on the previous page can be shortened by statements using numbers.

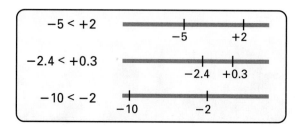 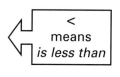

Instead of "earlier," "lower," and "colder," you can use the more general word *less*. So −5 < +2 can be read as −5 is **less than** +2. You can also say: +2 is greater than −5, +0.3 is greater than −2.4, and −2 is greater than −10. The short notation is:

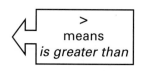

5. Write words for each of these statements.

   a. +7 > −7   c. −10 < +9
   b. −6 < −5$\frac{3}{4}$ d. −1000 > −2000

6. Make true statements using < and >.

   a. 789 ___ 798   c. +12 ___ −24
   b. −3.7 ___ −4.3   d. $\frac{1}{2}$ ___ $\frac{1}{3}$

To help see how the numbers are related, mathematicians use a number line that can be extended in both directions as far as you want! This is shown by the two arrows.

If the number line is horizontal, the *positive* numbers are on the *right* of 0 and the *negative* numbers are on the *left* of 0.

Walking Along the Number Line

Often the positive numbers are written without the +sign, like this.

If you move along the line in the positive direction, the numbers that you pass become larger.

If you move in the negative direction, the numbers become smaller. The movement from 4 to −6 is in the negative direction, so −6 < 4.

7. **a.** The distance between 4 and −6 is equal to 10. How can you explain that?

   b. What is the distance between 14 and −16?

Look at the curved number line.

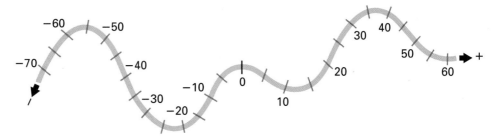

The difference between 60 and 20 is 40. That is just the distance on the number line!

8. **a.** What is the difference between 45 and −10?

   b. What is the difference between −15 and −65?

9. **a.** Give three pairs of numbers, each consisting of a positive and a negative number, with a difference of 100.

   b. Give three pairs of negative numbers with a difference of 50.

B Walking Along the Number Line

Ronnie the Robot

We can move Ronnie along the number line by giving him an instruction:

- with one of the two words "ADD" or "SUBTRACT";
- followed by a positive or a negative number.

When the instruction begins with ADD, Ronnie looks in the positive direction.

If the number is positive, he moves forward.

If the number is negative, he moves backward.

Here are two examples:

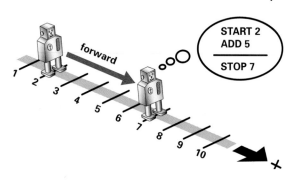
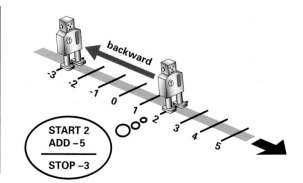

Suppose Ronnie is standing on the number 2.

The instruction is ADD 5. Ronnie looks in the positive direction and moves forward to the number 7. (See the first picture).

In the second picture, the instruction is ADD −5. Now he moves backward and stops at the number −3.

10. Ronnie starts at the number 2 each time.
    a. Where will he stop if the instruction is ADD 18?
    b. Where will he stop if the instruction is ADD −18?

11. Now Ronnie starts at the number −5 each time.
    a. Where will he stop if the instruction is ADD 5?
    b. Where will he stop if the instruction is ADD −5?

Walking Along the Number Line

Suppose the instruction starts with the word SUBTRACT.

Because of the word SUBTRACT, Ronnie now looks in the negative direction, as you see in the pictures, and:

- if the number is positive, he moves forward.
- if the number is negative, he moves backward.

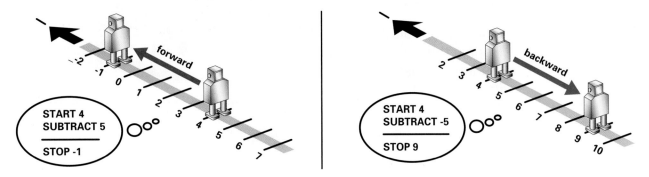

In the pictures, you see that Ronnie starts at the number 4.

So if the instruction is SUBTRACT 5, he stops at −1.

12 a. If the instruction SUBTRACT 5 is repeated, where does Ronnie stop this time?

 b. Where will he stop if the starting point is −4 and the instruction is SUBTRACT −4?

Ronnie is standing at the number −8. You want him to move forward to the number +8.

13. a. What instruction will you give?

 b. But Ronnie wants to move backward! Now, what instruction can you give to have him end at 8?

Now Ronnie is standing at the number −14, and you want to send him to −24.

14. What instruction will you give to have him stop there?

Activity

The Robot Game

To play this game, you need a number line on the floor.

You can make this by using colored tape or a rope or long string and cards with positive and negative numbers.

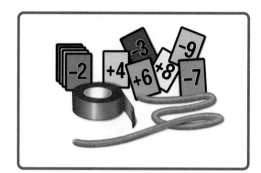

One student plays the role of Ronnie the Robot. Another student chooses from four instructions to move Ronnie.

ADD (+ ...)

ADD (– ...)

SUBTRACT (+ ...)

SUBTRACT (– ...)

Ronnie chooses a starting point. The second student gives an instruction with a number, and Ronnie moves along the line.

The student should give one of each type of instruction in any order. Other students check to see if Ronnie stops on the right spot.

After four moves, the game continues with two other students.

Walking Along the Number Line

15. Complete the following series of instructions. You can use the number line to support your thinking.

At the end of the chain, you can choose your own instructions.

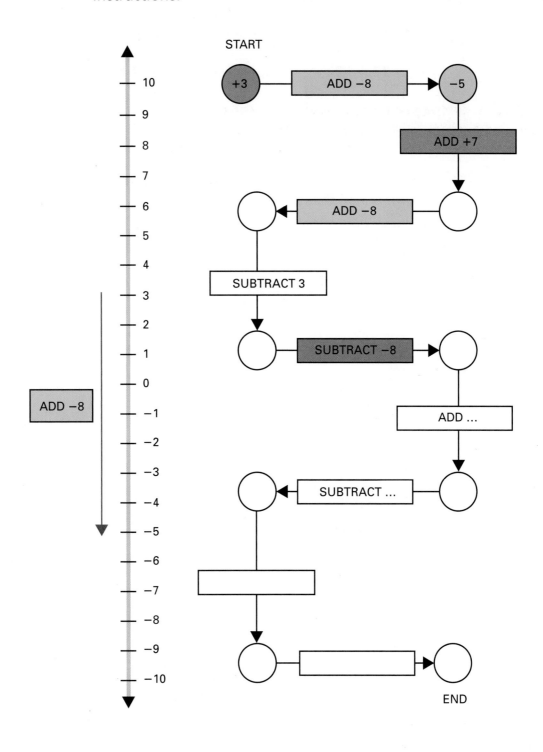

B Walking Along the Number Line

Summary

Positive and negative numbers can be ordered on a number line. The farther the number is to the right, the larger the number is. 17 is less than 28, but −17 is greater than −28.

The *difference* between two numbers is the distance between these numbers on the number line; for instance, the difference between 28 and −17 is equal to 45.

Movements along the number line can be indicated through instructions with ADD or SUBTRACT.

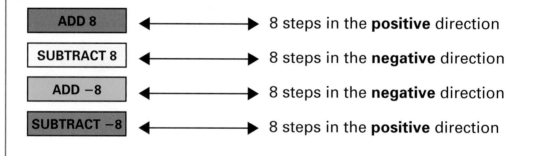

ADD 8	⟷	8 steps in the **positive** direction
SUBTRACT 8	⟷	8 steps in the **negative** direction
ADD −8	⟷	8 steps in the **negative** direction
SUBTRACT −8	⟷	8 steps in the **positive** direction

Check Your Work

1. **a.** Here you see a part of a number line with numbers ranging from −60 to 60. Fill in the blanks.

   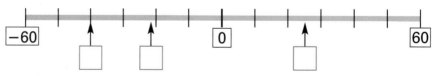

   b. Put one new positive and one new negative number on the number line yourself so that the difference between these numbers is 75.

2. Make true statements using <, =, or > and write each statement in words.

   a. −24 ___ 14

   b. −2000 ___ 2000

   c. −101 ___ −100

   d. $\frac{1}{4}$ ___ $\frac{1}{5}$

3. a. Ronnie the Robot starts at number 6. The instruction is ADD −9. Where does Ronnie stop?

   b. Write three different instructions for Ronnie. Use ADD as well as SUBTRACT.

4. Complete the following lines.

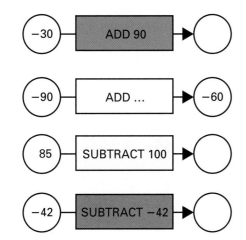

For Further Reflection

Explain why subtracting −8 is the same as moving 8 steps in the positive direction on a number line.

C Calculating with Positive and Negative Numbers

Adding and Subtracting

Ms. Parker is working in a building that has 40 stories above the ground floor. The building has an underground parking garage with 6 floors. These floors are indicated with negative numbers: −1 through −6.

The ground floor is indicated by 0, and the upper floors have numbers from 1 to 40.

Ms. Parker leaves her car at level −4 and enters the elevator. Then she rides the elevator for 28 floors.

1. At what level does she arrive?

2. The calculation in problem 1 can be shown as an addition.

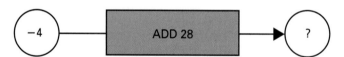

   a. Which movement of the elevator corresponds with the following?

   What is the result of the addition?

   b. Which movement of the elevator corresponds with the following?

   What is the result of the addition?

22 Operations

Calculating with Positive and Negative Numbers

3. **a.** What is the maximum number of levels that can be covered going up in the elevator?

   b. Write a statement using addition that uses this maximum number of levels.

   c. Now write a statement using addition that uses the maximum movement down.

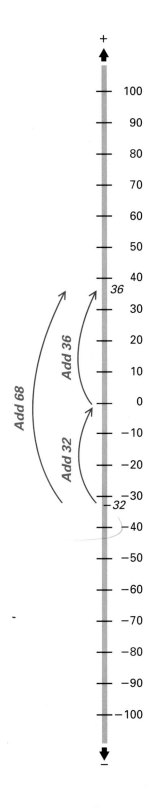

Suppose you wanted to do the calculation below. One way is to split the number 68 in two parts, as you can see in the number line to the left. The result is 36.

4. **a.** How does splitting 68 in this way help you to do the calculation?

   b. How can you change the picture to show the following?

From now on, such calculations will be written in a shorter way.

$$(-32) + (68) = (36)$$

$$(32) + (-68) = (-36)$$

Section C: Calculating with Positive and Negative Numbers 23

Calculating with Positive and Negative Numbers

5. Complete the following calculations. You may draw a number line if it is helpful.

   a. (30) + (−60) = ...

   b. (32) + (−58) = ...

   c. (32) + (−48) = ...

   d. (24) + (−48) = ...

   e. (48) + (−24) = ...

   f. (−30) + (60) = ...

   g. (−30) + (−60) = ...

   h. (−32) + (35) = ...

   i. (−32) + (28) = ...

   j. (−32) + (−28) = ...

24 Operations

Calculating with Positive and Negative Numbers

6. Use **Student Activity Sheet 2** to complete the two "adding trees."

a.

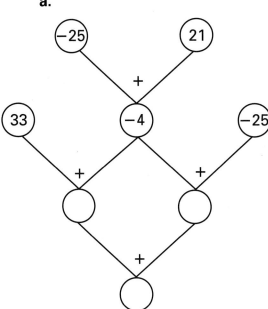

b.

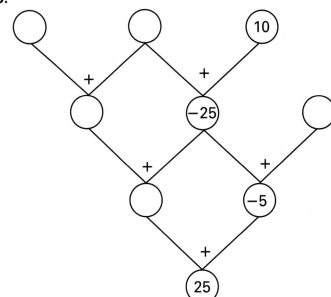

7. There are many ways to fill in numbers in the following adding tree. Copy the table and fill it in so that all of the numbers are different and the final result is 0.

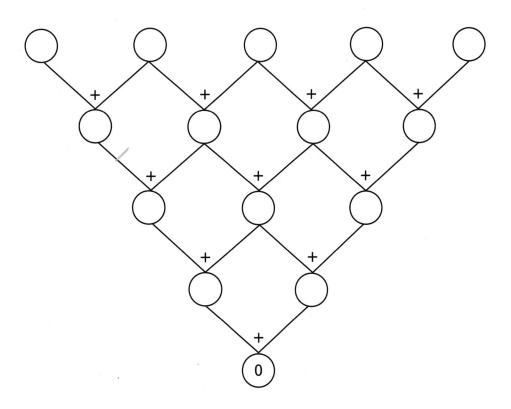

Section C: Calculating with Positive and Negative Numbers 25

Calculating with Positive and Negative Numbers

Recall the building in problem 1.

The distance between the levels 24 and −5 is 29 stories. This is written as:

(24) − (−5) = (29)

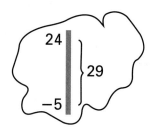

This is an example of a subtraction.

8. a. Complete the following subtractions.

i. (14) − (−5) = ( )

ii. (4) − (−5) = ( )

iii. (−4) − (−5) = ( )

b. Compare these calculations to:

i. (14) + (5) = ( )

ii. (4) + (5) = ( )

iii. (−4) + (5) = ( )

In general, *subtracting* −5 gives the same result as *adding* 5.

9. Give three examples that show that subtracting −10 gives the same result as adding 10.

10. Complete the following statement: *Subtracting 8 gives the same result as adding ___.*

Calculating with Positive and Negative Numbers

11. Complete the calculations.

a. $8 - (-10) =$
   $8 + 10 =$

b. $-8 + (-20) =$
   $-8 - =$

c. $-22 + 12 =$
   $-22 - =$

d. $75 - = 99$
   $75 + = 99$

Activity

The Integer Game

Here are rules for a card game using 40 index cards. Number 10 cards with consecutive integers from 1 to 10, using a black marker, with one number on each card. Repeat with 10 cards using a red marker, numbering each card with consecutive **integers** from -10 to -1. Make two complete sets of each color. Play with two or three players.

- Shuffle all of the marked cards together.
- Each person gets 7 cards and the rest go face down in a stack in the middle of the table.
- Decide who plays first. The first player uses as many cards as possible that add up to -2 and lays them on the table for others to see.
- Put the cards used with -2 on the bottom of the stack and turn over a new card from the stack.
- The first player tries to use cards that add up to the number on the new card. If that is not possible, it is the next player's turn.
- If the player does not make the number the first time, the player draws a card from the stack. If the player still cannot make the number, it is the next player's turn.
- The first player without any cards wins the game.

Section C: Calculating with Positive and Negative Numbers 27

Calculating with Positive and Negative Numbers

Temperatures and Altitudes

During the winter, Karen's class was given the assignment to record the high and low temperatures for one week.

Here are the temperatures Karen recorded.

	Sun	Mon	Tue	Wed	Thu	Fri	Sat
High Temp (°C)	9	3	4	−2	−1	3	−1
Low Temp (°C)	−1	0	−4	−3	−7	−1	−2

12. **a.** Which day was the coldest? Which day was the warmest? Explain your answers.

    b. Write the low temperatures in order from coldest to warmest. Was the temperature below freezing every day?

13. Calculate the **mean** high temperature for the week. What was the mean low temperature? Show your calculations.

28 Operations

Calculating with Positive and Negative Numbers

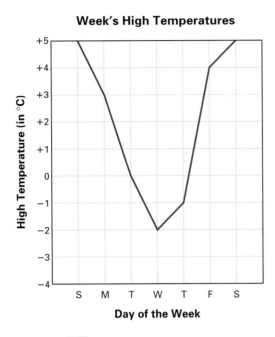

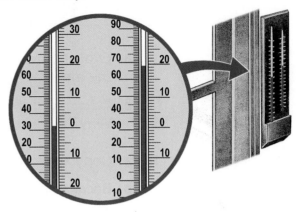

The next week, Karen again kept records of high and low temperatures. Instead of writing them in a table, she made a graph. This graph shows the high temperature for each day.

14. Reflect Compare the high temperatures from this graph to those from the table for the previous week. Which week was warmer? Give reasons to support your answer.

Karen calculated the mean low temperature in the second week to be −1°C.

15. Come up with one possibility for each of the seven daily low temperatures that week in degrees Celsius.

Here is a list of high temperatures for the whole month of January.

Sun	Mon	Tue	Wed	Thu	Fri	Sat
	−1	+4	0	−1	−1	+1
−1	+3	0	0	−1	+1	+1
−3	−3	−5	−1	0	0	+1
+1	+2	+2	0	−2	−2	−1
−1	+2	+2	−2			

Karen and her classmate Diego both want to calculate the mean high temperature for January. Diego starts crossing out pairs of opposite numbers (for example, +2 and −2) and then adds all of the remaining temperature numbers.

16. By what number does Diego divide to find the mean high temperature?

Section C: Calculating with Positive and Negative Numbers 29

Calculating with Positive and Negative Numbers

Karen starts by tallying how often each temperature occurred.

−5	−4	−3	−2	−1	0	+1	+2	+3	+4	+5
				///	/	/			/	

17. a. Explain how Karen should continue in order to calculate the mean.

b. Reflect Which method do you like better, Karen's or Diego's? Why?

18. a. Copy and finish Karen's tally table.

b. Because there are three days with a temperature of −2°C, Karen calculates:

$$3 \times -2 = -2 + -2 + -2 = \underline{}$$

What is the answer? Make calculations like this for each entry in the table.

c. Calculate the mean high temperature.

Higher and Higher

Karen and Diego's school is near a ski area. There are three ski lifts on the mountain. The lowest one is 1,300 m above sea level. From this lift station, a ski lift can take you to the highest lift station at 2,300 m. There is a middle lift station halfway between these two stations.

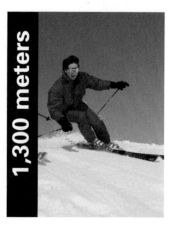

Calculating with Positive and Negative Numbers

A sign at the lower lift station shows the temperatures at each lift station at different times of the day.

Altitude (in meters)	Ski-Lift Station	Temperature (C°) at 8:00 A.M.	Noon	6:00 P.M.
2,300	Upper	−8		
1,800	Middle			
1,300	Lower	−2		

19. a. What do you think the temperature at the middle lift station was at 8:00 A.M.?

b. What do you notice about the temperature as you move higher up the mountain?

At noon, the temperatures at all three ski lift stations have risen by 5 degrees.

20. a. Write down the noon temperature for each ski lift station.

b. Assume that the differences in temperature between the ski lift stations are always the same. What might the temperatures be at 6:00 P.M.?

In the mountains, a change in height results in a regular change in temperature.

21. In this ski area, what happens to the temperature each time you move up 500 m?

Calculating with Positive and Negative Numbers

The next day at noon, the temperature at the middle lift station is −1°C.

22. At noon, what temperature will the sign show for each of the following places?

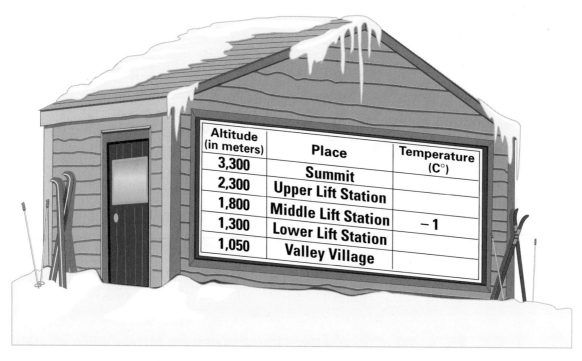

Altitude (in meters)	Place	Temperature (C°)
3,300	Summit	
2,300	Upper Lift Station	
1,800	Middle Lift Station	−1
1,300	Lower Lift Station	
1,050	Valley Village	

Here are the high temperatures for a week at the middle lift station.

	Sun	Mon	Tue	Wed	Thu	Fri	Sat
High Temp (°C)	−5	−3	−3	0	−4	−3	−5

23. What was the mean high temperature that week at the upper lift station? What was it at the lower lift station?

The general rule you discovered between temperature and altitude tends to be true all over the world: For each 500 m you go up, the temperature drops by about 3°C.

24. a. What happens to the temperature if you move up 1,000 m? 250 m? 100 m?

b. Write a rule describing what happens to the temperature as you go down.

32 Operations

Calculating with Positive and Negative Numbers

You may find heavy snow at higher altitudes. Sometimes snow remains at high altitudes all summer long.

Suppose that the summit of a mountain is 4,418 m above sea level. This mountain descends into the sea. At sea level, the temperature is 23°C.

25. Is the snow melting at the summit? You may continue to build a table like the one below to help you.

Altitude (in m)	0 (sea level)	500	1,000	
Temperature	23°C			

Marcos is working on the following problem.

The temperature at the foot of a mountain is 12°C. What would the approximate temperature be at 300 m up the mountain?

Marcos writes 12 + 3 × −0.6 =

26. Explain how this calculation fits the problem. Then find the answer.

Calculating with Positive and Negative Numbers

Summary

In this section, you found that *subtracting* −15 has the same result as *adding* 15.

$$18 - (-15) = 33$$
$$18 + 15 = 33$$

You also found that *subtracting* 15 has the same result as *adding* −15.

$$18 - 15 = 3$$
$$18 + (-15) = 3$$

If you write these calculations without circling the numbers, use parentheses for the negative numbers: 18 − (−15) = 33 and 18 + (−15) = 3. Note: The command about using parentheses may lead to confusion later if they are not used.

A repeated addition can be written as a multiplication; for instance:

$$(-2) + (-2) + (-2) + (-2) + (-2) = 5 \times (-2) = -10$$

A general rule describing the relationship between temperature and altitude:

For each 500 m you go up, the temperature drops by about 3°C.

Check Your Work

1. The following tree uses *adding* and *subtracting*. If the sign is −, you have to subtract the number above the sign on the right from the number above the sign on the left. Copy and complete the tree.

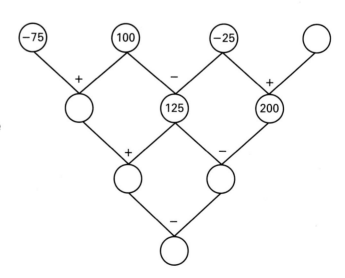

34 Operations

2. For ten days, the high temperatures were recorded. On four days, the high temperature was −3°C; on three days, it was −2°C; on one day, it was −1°C; and on two days, it was +2°C.

   a. Explain why you can begin to find the mean temperature as follows:

   $$(4 \times -3) + (3 \times -2) + (1 \times -1) + (2 \times 2) =$$

   b. Finish the calculation for the mean temperature.

3. Which of these four statements will *always* be true? Explain your answer.

   a. A positive number added to a positive number results in a positive number.

   b. A negative number added to a negative number results in a negative number.

   c. A negative number added to a positive number results in a positive number.

   d. A negative number subtracted from a positive number results in a positive number.

For Further Reflection

Describe how multiplication is related to addition and how this can help you understand how to operate with negative numbers.

Adding and Multiplying

Calculations Using Differences

The scores (in %) on a math test for a group of 20 students are:

74	74	76	80	84
84	84	84	85	88
88	91	91	93	93
93	96	96	97	99

The teacher guesses that the mean will not be far from 85.

To check this, he calculates the difference between each score and 85.

He considers this difference negative if the score is less than 85 and positive if the score is more than 85.

Adding and Multiplying

Here is a beginning of the list of differences.

−11 −11 ....

....

....

.... +14

1. a. Complete the list and add all of the numbers.

   b. Do you think the mean is less than or more than 85? Explain your thinking.

   c. Divide the sum of all of the differences from your list by 20. How can you use this result to find the mean score of the group?

The teacher asked Iris, the student with the highest score, to also calculate the mean.

He did not tell her his result.

Iris guessed the mean would be 90, so she made a list of differences from 90 and calculated the mean score by using the mean of the differences.

2. Make Iris's list of differences and use this list to calculate the mean. Did you find the same result as in problem 1c?

D Adding and Multiplying

Multiplication with Positive and Negative Numbers

4 times 6 means 6 + 6 + 6 + 6, and the result is 24.

What does *4 times* −2 mean?

In Section C, you met Diego, who calculated:

$$4 \times -2 = -2 + -2 + -2 + -2 = -8$$

In the list of differences (see problem 2), you found 2 times −16, 4 times −6, and 2 times −2.

3. Use these numbers to write three calculations like Diego did with 4 × − 2.

You have seen some examples of multiplying a positive times a negative. But how would you think about −4 times 6 or, even worse, −4 times −6? What can be the meaning of −4 *times* something?

In mathematics, it is possible to do multiplications for those two examples. First look at the half number line.

The line has a double scale. The bottom numbers are multiples of 6.

The picture shows, for instance:

$$4 \times 6 = 24$$

4. Suppose that this line is continued to the right. Which number will be just above 84? Above 420? Write the corresponding multiplication.

Now continue the double number line to the left.

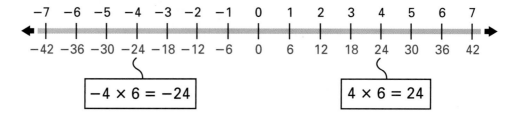

Adding and Multiplying

The negative numbers are called *negative multiples* of 6.

5. a. Reflect How can you explain that $-10 \times 6 = -60$?

   b. Do you think 6×-10 will have the same result? Why or why not?

6. a. Make a double number line in such a way that the black 1 corresponds to the red 8.

   b. Write three multiplications of the form .... × 8 = ..., using negative numbers.

7. a. Complete the multiplication table for 11.

   b. What does the −11 by the arrow in the table have to do with these multiplications?

$3 \times 11 = 33$
$2 \times 11 = 22$ ⎞ −11
$1 \times 11 = 11$ ⎠ −...
$0 \times 11 = 0$ ⎠ −...
$-1 \times 11 = ...$ ⎠ −...
$-2 \times 11 = ...$ ⎠ −...
$-3 \times 11 = ...$

Now consider the multiplication of negative numbers.

Use a double number line. The numbers are multiples of −6.

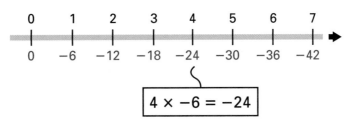

8. a. First look at the number line above. Which number will be below 8? Below 12? Write the multiplication statement for each of these.

   b. Which number will be above −60? Above −300?

Section D: Adding and Multiplying 39

D Adding and Multiplying

In the number line below, you see the continuation of the numbers to the left.

Reading from left to right, the numbers below the line are going down, but the numbers above the line are going up.

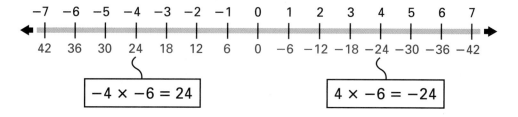

So you can see that the negative multiples of −6 are positive!

9. a. Which number corresponds to −9? Write the multiplication statement.

   b. Which number corresponds to 66? Write the multiplication statement.

10. a. Complete the multiplication table shown below for −8.

   b. What does the +8 by the arrow in the table have to do with these multiplications?

$$3 \times -8 = -24$$
$$2 \times -8 = -16$$
$$1 \times -8 = -8$$
$$0 \times -8 = 0$$
$$-1 \times -8 = \ldots$$
$$-2 \times -8 = \ldots$$
$$-3 \times -8 = \ldots$$

+8
+....
+....
+....
+....

11. Complete the following calculations.

20 × −15 =	30 × 5 =	10 × 25 =
20 × 15 =	30 × −5 =	0 × 15 =
−20 × 15 =	30 × −15 =	−10 × 5 =
−20 × −15 =	30 × −25 =	−20 × −5 =

40 Operations

Use **Student Activity Sheet 3** for problems 12 and 13.

12. Complete the multiplication tree.

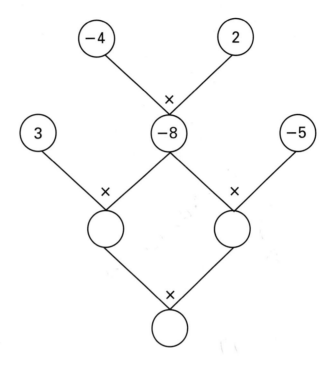

13. Here you see a tree with different kinds of operations (+ and x). Complete this tree.

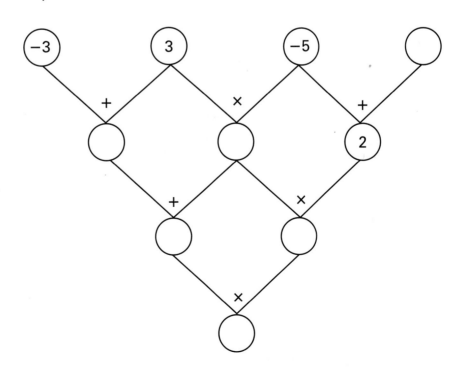

Section D: Adding and Multiplying 41

D Adding and Multiplying

Summary

Multiplication with a positive whole number is the same as a repeated addition; for instance:

$5 \times 7 = 7 + 7 + 7 + 7 + 7 = 35$

$5 \times (-7) = (-7) + (-7) + (-7) + (-7) + (-7) = -35$

With the help of a pattern on the number line, you can find the results of multiplying by a negative whole number; for instance:

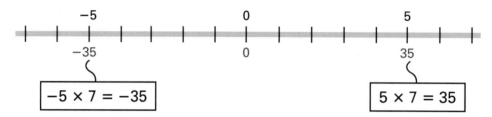

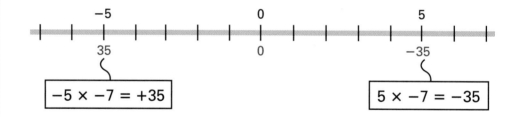

There are four rules for multiplication of integers.

positive × *positive* = *positive*

positive × *negative* = *negative*

negative × *positive* = *negative*

negative × *negative* = *positive*

Check Your Work

1. Copy and complete the multiplication tree.

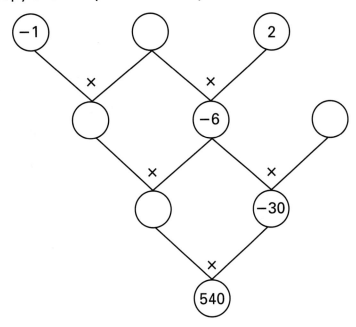

2. Calculate the mean score of a test with the following results. Use a table of differences.

66	68	74	75	75	75
77	77	77	79	80	81
82	83	83	83	85	85
91	95	97	98	100	100

3. Use all three of the operations adding, subtracting, and multiplying and at least two negative numbers to make a calculation string for each of the numbers from −1 to −10 (for example, (−2 × 5) − (−4) + 5 = −1).
   Have a classmate check your answers.

 ## For Further Reflection

Write a letter to a friend explaining the rules for multiplication of positive and negative numbers.

Section D: Adding and Multiplying 43

Operations and Coordinates

Directions

Diego invites some friends to his home for a sleepover. "It's easy to find," he says. "From Pioneer Park, you go three blocks east and then two blocks south. There is my house."

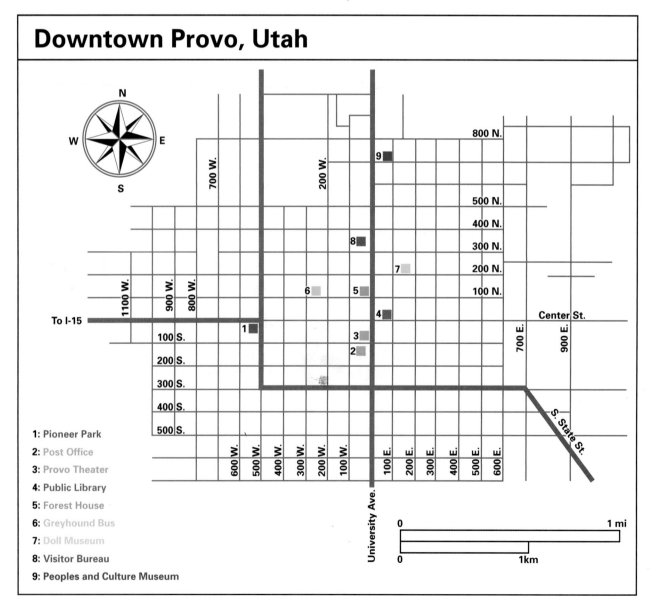

Downtown Provo, Utah

1: Pioneer Park
2: Post Office
3: Provo Theater
4: Public Library
5: Forest House
6: Greyhound Bus
7: Doll Museum
8: Visitor Bureau
9: Peoples and Culture Museum

Operations and Coordinates

1. The directions Diego gave can be noted as (E 3, S 2).

   a. Use the same notation to give directions from Pioneer Park to the Peoples and Culture Museum.

   b. Use the same notation to give directions from the Public Library to the Peoples and Culture Museum. Start with the direction for east or west and then state the direction for north or south.

Now that Diego has learned about positive and negative numbers, he decides to change his system. "If Pioneer Park is my starting point, I can use positive numbers for going east and negative numbers for going west, and then positive numbers for going north and negative numbers for going south—just like using two number lines that are perpendicular."

Here is Diego's sketch.

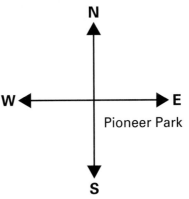

2. Use the new notation to give directions from Pioneer Park to Diego's house.

3. a. What directions does (+2, −1) mean in Diego's notation?

   b. What does (0, 0) mean in Diego's notation?

4. Could you use (−8, −5) to give directions in your town? Explain why or why not.

Section E: Operations and Coordinates 45

Operations and Coordinates

Changing Shapes

To communicate about the locations of points on a grid, it is useful if everybody uses the same language and notation. You need to choose a starting point first.

Mathematicians and scientists use a grid with a starting point called the **origin** and a **horizontal** and **vertical axis** with numbers very much like horizontal and vertical number lines. This type of grid is called a **coordinate system**. Positive and negative numbers on the axes can be extended as far as you wish.

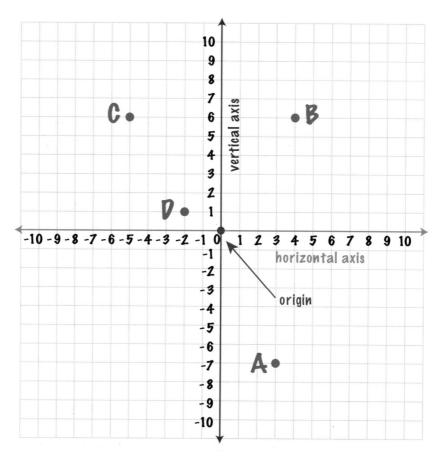

On the coordinate system, the **coordinates** for point A are written as (3, −7). Note that you start with the horizontal direction and start counting at (0, 0), the *origin*. The + sign for a positive number is usually not shown; it is understood.

5. a. How would you describe the location of points B and C?

   b. Why does it make sense to use the coordinates (0, 0) for the origin?

6. Is (−2, 1) the same point as (1, −2)? Explain your reasoning.

Operations and Coordinates

7. Draw your own coordinate system. Use graph paper.

    a. Draw vertical and horizontal axes at least 10 units long. Put a number scale on each axis and a 0 for the origin.

    b. Plot the following points in your coordinate system.

    (1, 1), (3, 3), (2, −1), (−2, −1), (−3, 1), (−4, 0), (−3, 2),

    (−2, 2), (−1, 1)

    Connect the points in order starting and ending at (1, 1). Add an eye to complete your drawing.

Use Student Activity Sheet 4 for problems 8–12.

8. a. Plot the following points on your coordinate system. Remember that the first coordinate of the pair names a position going right or left in the horizontal direction, and the second coordinate names a position going up or down in the vertical direction.

    (1,1), (5,1), (6,2), (7,2), (7,1), (8,1), (9,2),

    (9,4), (7,4), (6,5), (5,5), (1,3), (0,3), (1,1)

    b. Connect the points in the order they are shown in 8a. What is the result?

For problems 9 and 10, predict what you think will happen to the drawing you made for problem 8. Check your prediction by making a new drawing. For each problem, start with the drawing you made for problem 8.

9. Add −10 to the first coordinate of each point. What happens?

10. Add 2 to the first coordinate and add −5 to the second coordinate of each point. What happens?

11. What should you do to the coordinates if you want to move the drawing up three units and to the right five units?

12. What should you do to the coordinates if you want the drawing to become twice as large?

Operations and Coordinates

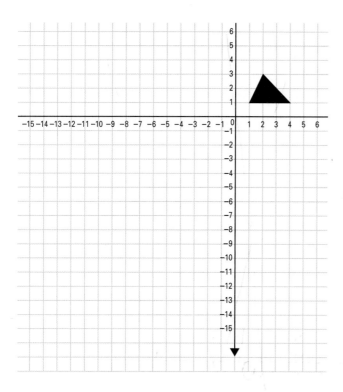

13. On a new copy of **Student Activity Sheet 4,** draw the triangle shown on the graph. Only part of the graph is shown on the left.

    a. What are the coordinates of each **vertex**?

    b. Multiply all of the coordinates by −3. What are the new coordinates of the vertices?

    c. Draw the new figure in the same coordinate system. Describe how the new figure is related to the original one.

Carrie wonders what would happen to the figure she made in problem 13 if she multiplied the coordinates by −3 a second time. This is what some of her classmates think.

John says, "It would be upside down and three times as big."
Mauri says, "I guess it would be nine times as big."
Taye says, "I think it would become the first figure."
Emily says, "The coordinates of the top point would be (9, 8)."

14. a. **Reflect** Comment on the thinking of each of Carrie's four classmates.

    b. Write down everything you know about what will happen to the figure.

15. a. What would happen to the triangle you drew for problem 13a if you multiplied only the first coordinate by −3 and kept the second coordinate the same?

    b. What would happen to the triangle you drew for problem 13a if you kept the first coordinate the same and multiplied only the second coordinate by −3?

Operations and Coordinates

Gil, Lashonda, and Greg are discussing how they might shrink a triangle.

Gil says, "You could multiply the coordinates by −2."

Lashonda says, "That is not right. You would have to multiply the coordinates by $\frac{1}{2}$."

Greg says, "Why not multiply by $-\frac{1}{2}$?"

16. Which of these statements do you think is/are correct?

Three figures have been drawn in the coordinate system below. Use a new copy of **Student Activity Sheet 4** for problems 17–19.

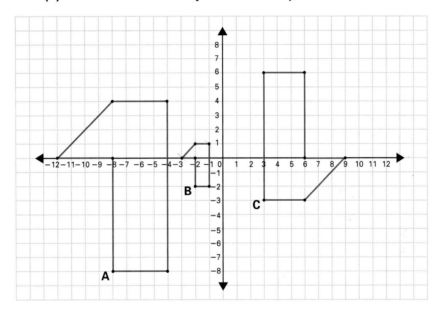

17. a. Choose one of the figures A, B, or C. Describe how you can get the other two figures from the one you chose by multiplying or dividing the coordinates of all of the points of that figure.

   b. Check your answer and show how you checked it.

For problems 18 and 19, you will multiply a figure by a number. This means that you will multiply the coordinates of all of the points by that number.

18. Start with figure B and multiply figure B by −1. Draw the new figure on another copy of **Student Activity Sheet 4** and name it D.

19. Multiply figure D by −1. What do you notice? What does this tell you about −1 × −1?

Section E: Operations and Coordinates 49

Operations and Coordinates

Summary

To locate points on a grid, you can use a coordinate system. The origin is the point at the intersection of the horizontal axis and the vertical axis, which are like perpendicular number lines. Each point can be located by using two coordinates that tell where the point is located relative to the origin. Positive directions are up and to the right; negative directions are down and to the left.

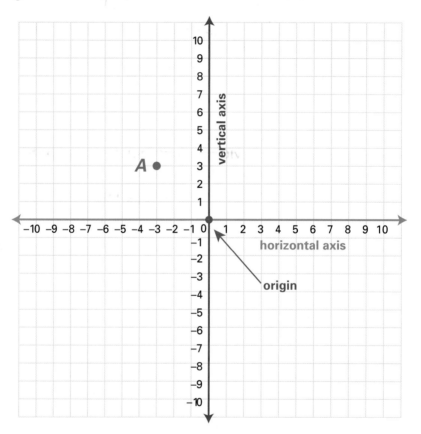

You can change the position of a figure drawn in a coordinate system by adding a number to or subtracting a number from each coordinate, or by multiplying or dividing the coordinates by some number. You need to remember the rules for these operations in order to do this.

Check Your Work

1. a. Draw your own coordinate system on grid paper by making vertical and horizontal axes 14 units long. Put a number scale on each axis, starting with 0 at the origin. Plot the following points and connect the points with line segments in this order.

    (−5, 0), (−2½ , 1), (−1, 1), (−1, 3), (1, 6), (1, 3), (2, 1), (7, 1), (5, 0)

    b. Use the horizontal axis as a mirror to draw the other half of the picture of a bird in flight.

    c. Write the coordinates of the points you drew for problem 1c.

2. Do you think the following statement is always true? Give an explanation for your answer.

    "If you multiply or divide the coordinates of a figure by a number, the size will always change."

3. Suppose a figure called A is multiplied by +2 and called figure B. Then figure B is multiplied by −1 and called figure C, and figure C is multiplied by +½ and called figure D. How can you get figure D directly from figure A?

For Further Reflection

How can you make a figure smaller by multiplying?

Additional Practice

Section A Positive and Negative

Sharon Taylor is a salesperson for a toy manufacturing company. To sell her company's toys to different stores, she must travel quite often. Sharon is flying from New York to Seattle for a meeting with a retail chain.

1. Sharon's plane left New York at 3:00 P.M., and it is a six-hour flight from New York to Seattle. What time will she arrive in Seattle?

After her trip to Seattle, Sharon returns to New York. She wants to schedule a conference call by telephone with an Italian toy retailer in Rome. Sharon works between 9:00 A.M. and 5:00 P.M. each day. The Italian retailers also work from 9:00 A.M. to 5:00 P.M.

2. When can Sharon schedule a conference call?

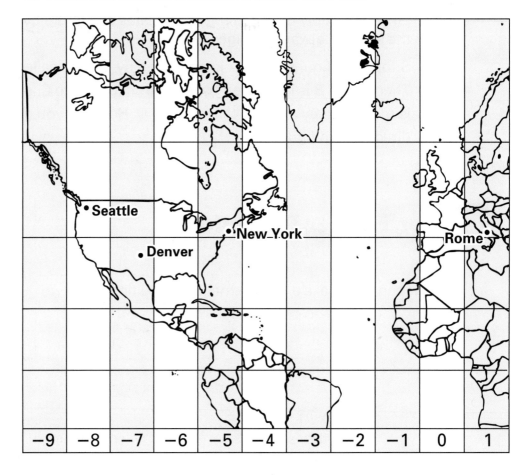

52 Operations

A group is planning a hiking trip from point A to point G on the trail mapped below.

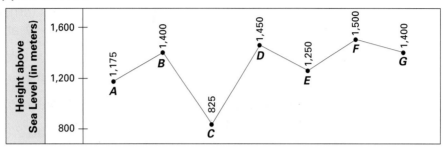

Someone in the group says, "Wow, we are going to be only a few hundred meters higher when we end. That does not sound like a difficult hike." As the hiking guide, you decide it would be helpful to explain how much climbing they will have to do.

3. **a.** How many meters is the difference in height between point A and point B? How many meters does the group descend between points B and C?

   b. Complete the table, which shows how many meters the hikers will have to ascend and descend. Use + for going up and – for going down.

	Height (m)	Up	Down	
A	1,175			
B	1,400	+225		
C	825		–575	
D	1,450			
E	1,250			
F	1,500			
G	1,400			
Total		+	–	=

   c. What are the totals in the columns "Up" and "Down"? What does this mean for the difference in height between points A and G?

   d. Use the table you made for problem 3b to decide if this is a difficult hike. Give mathematical reasons to support your answer. If you want to, you may use the general rule: There are about 3 ft in 1 m.

Additional Practice 53

Additional Practice

Section B Walking Along the Number Line

Draw a number line exactly 14 centimeters (cm) long. Use your ruler. Put −7 on the left end of the number line, +7 at the right end, and 0 in the middle.

1. **a.** Use an arrow to indicate −4 on your number line.

   b. Use an arrow to indicate 2.8 on your number line.

   c. What is the distance between −4 and 2.8?

2. The lowest point in Washington, D.C., is +0.3 m, and the lowest point in Louisiana is −2.4 m. What is the difference in height between these two points?

3. **a.** Use a short notation to write "negative seven is less than positive two."

   b. Make true statements using < or > or =.

   +75 ____ +57   −100 ____ +10 $5\frac{3}{4}$ ____ 5.75

   −3 ____ +3   −100 ____ −1000 $-2\frac{1}{2}$ ____ $-2\frac{2}{5}$

4. The temperature in Bergen, Norway, was −8 degrees Celsius overnight. During the daytime, the temperature went up to a maximum of +4 degrees Celsius. How much did the temperature rise?

5. Complete the following.

   a. −25 ADD 17 → ____   **d.** 25 SUBTRACT −25 → ____

   b. 25 ADD −17 → ____   **e.** −25 SUBTRACT ____ → 50

   c. 25 SUBTRACT 25 → ____

Section C Calculating with Positive and Negative Numbers

1. Fill in the blanks.

   a. Adding 8 gives the same result as subtracting ____.

   b. Subtracting 10 gives the same result as adding ____.

   c. 15 − (−3) = ____

   d. 15 + ____ = 18

Additional Practice

Here is a list of high temperatures in degrees Celsius for one week at a ski lift station.

Day	Sun	Mon	Tue	Wed	Thu	Fri	Sat
High Temperature (°C)	−7	−5	−5	−3	−5	−7	−6

2. Find the mean high temperature for that week at the station. Show your work.

3. The following tree uses addition and subtraction. Copy and fill in the tree going from left to right.

   If the sign is negative (−), you have to subtract the number on the right from the number on the left.

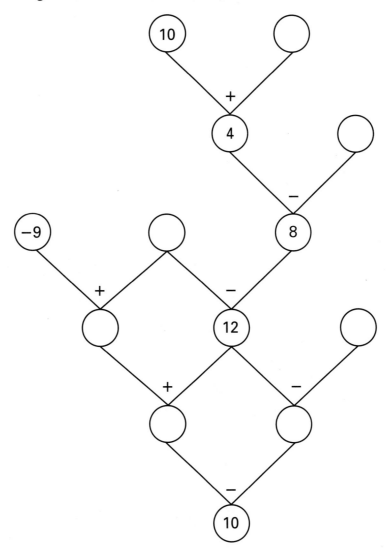

Additional Practice

Section D Adding and Multiplying

1. Are the following statements always true? If not, show an example for which the statement is not true.

   a. A positive number multiplied by a positive number gives a positive number.

   b. A positive number added to a negative number gives a positive number.

   c. A negative number multiplied by a negative number gives a positive number.

The students in Ms. Makuluni's class have measured their pulse rate for half a minute while sitting at their desks. Here are the results.

34	35	35	36	35	36
33	37	32	34	36	34
30	37	34	38	33	35
31	35	36	33	38	37

2. a. Make a list of positive and negative differences from 35.

   b. Is the mean pulse rate, measured for half a minute, more or less than 35? Show your work.

3. Complete the calculations in both tables.

Table 1

+	−8	−5	−2	1	4.5
6					
2		−3			
−2					
−6					
−10					

Table 2

×	−8	−5	−2	1	4.5
6					
2					
−2					
−6		30			
−10					

4. Write five different calculations using positive and negative numbers.

   Fractions and decimal numbers are also allowed.
   You may use addition, subtraction, and multiplication.
   Don't make your problems too hard! You must include a list with correct answers for your calculations.

Additional Practice

Section E Operations and Coordinates

Use graph paper for problems 1–5.

1. a. Draw a coordinate system. Put number scales on it and mark 0 for the origin.

    b. Plot these points on your coordinate system: A (−1, −2), B (3, −2), C (4, 1), and D (0, 1).

    c. Connect the points in alphabetical order.

2. What are the new coordinates of your figure if you move it three spaces to the left and also three spaces up? Write the coordinates and draw the new figure in the same coordinate system you made for problem 1.

3. On **Student Activity Sheet 5,** you see a drawing of a letter F in a coordinate system.

    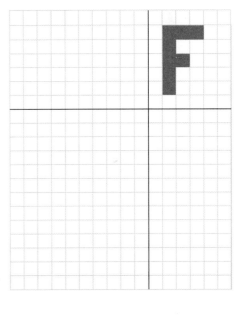

    a. Multiply each first coordinate of the letter F by −2 and keep each second coordinate the same. Draw the new figure in the same coordinate system and mark it with an A. Use a ruler.

    b. Keep each first coordinate of the original letter F the same and now multiply each second coordinate by −2. Draw the new figure in the coordinate system and mark it with a B.

    c. How did the shape of the original letter F change in parts **a** and **b**?

    d. What happens if you multiply both coordinates of the original letter F by −2? Make a drawing to support your reasoning.

Section A Positive and Negative

1. No, the trip back from Seattle did not take longer. Discuss your answer with a classmate. Sample explanation:

   The airline used local times for departure and arrival. There is a two-hour time difference between Seattle and Minneapolis. So 11:45 A.M. in Seattle is 1:45 P.M. in Minneapolis. The flight time was $3\frac{1}{2}$ hours both ways.

2. Some examples of situations where you can use positive and negative numbers:

   a. Below and above zero on a thermometer scale in degrees Celsius. Below zero is negative (−) and above zero is positive (+).

   b. Gain and lose yardage in football. Use (+) for gains and (−) for loss.

   c. Owing somebody money (−) or getting money (+).

   d. Height of the water in a lake, compared to a set level (0). Use (−) if the level has dropped below the set level and (+) if it has risen above the set level.

   e. Share your example with the whole class if it was not mentioned here.

3. a. The distance between the highest point (8,850 m) and the lowest point (−11,000 m) in the list is 19,850 m.

   b. No, you cannot use scale 1:100. Scale 1:100 means that 1 cm in the drawing equals 100 cm, or 1 m in reality. The length of the scale would be 19,850 cm long (or 198.5 m, which is really long!). Note that in a scale drawing, you can use negative numbers as well.

Answers to Check Your Work

4. a. You would end up lower than where you started. The total uphill (+) is less than the total downhill (−).

   b. You end up 30 m lower than the starting point. Sample strategy:

   Cancel out + 37 (uphill) and − 37 (downhill)
   + 230 (uphill) and − 130 (downhill) results in + 100
   + 100 and + 110 and + 140 results in + 350
   + 350 and − 340 results in + 10
   + 10 and − 40 results in −30

   Using a drawing to show how much you went uphill and downhill may help.

Section B Walking Along the Number Line

1. a. From left to right, the following numbers should be filled in: −40; −22; 25

   b. You can have many different answers. Discuss your answer with a classmate. One example: −45 and 30.

2. a. −24 < 14   negative twenty-four is less than fourteen

   b. −2000 < 2000   negative two thousand is less than two thousand (or 2000)

   c. −101 < −100   negative one hundred one is less than negative one hundred

   d. $\frac{1}{4} > \frac{1}{5}$ one fourth is greater than one fifth

Answers to Check Your Work

3. a. START 6
   ADD −9
   STOP −3

   b. Discuss your answers with a classmate. Sample answers:

   START −2   START −2
   ADD −8   SUBTRACT −4
   STOP −10   STOP 2.

4. Use a number line if it is helpful.

   −30 (ADD 90) → 60   85 (SUBTRACT 100) → −15

   −90 (ADD 30) → −60   −42 (SUBTRACT −42) → 0

Section C Calculating with Positive and Negative Numbers

1. Remember that if the sign is "−," you have to subtract the number above on the right from the number above on the left.

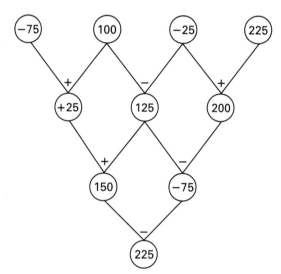

2. a. Different answers are possible. Sample responses:

   There were four −3's, three −2's, one −1, and two 2's. You should multiply each temperature by the number of times that temperature occurred and add the products. Then divide by 10 to find the mean temperature.

   b. (4 × −3) + (3 × −2) + (1 × −1) + (2 × 2) =
   $\qquad$ −12 + −6 + −1 + 4 = −15
   $\qquad\qquad$ −15 ÷ 10 = −1.5

   The mean temperature is −1.5°C.

60 Operations

Answers to Check Your Work

3. **a.** Always true. If you begin on the right and add a positive number, you move further to the right on the number line. Example: 30 + 60 = 90

   b. Always true. If you begin on the left and add a negative number, you move further to the left on the number line. Example: − 5 + (− 10) = − 15

   c. Not always true. Example: 6 + (− 10) = − 4

   Note that in case of a "not always true" statement, you need to give only one "counter-example" to show that the statement is not true.

   d. Always true. Subtracting a negative number gives the same result as adding a positive number. So you begin on the right and move farther to the right. Example: 30 − (− 20) = 50

Section D Adding and Multiplying

1.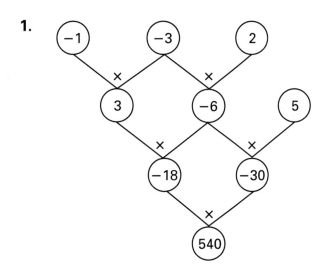

2. First make a list of differences from your estimated mean. Suppose you estimated the mean to be 80.

−14	−12	−6	−5	−5	−5
−3	−3	−3	−1	0	+1
+2	+3	+3	+3	+5	+5
+11	+15	+17	+18	+20	+20

Answers to Check Your Work 61

Answers to Check Your Work

Use any method to calculate the total of all differences: +66

You know now that the mean is above 80 since +66 is positive.

66 ÷ 24 = 2.75

The mean score is 80 + 2.75 = 82.75.

3. Have a classmate check your answers. If you do not agree, ask your teacher.

Section E Operations and Coordinates

1. a., b.

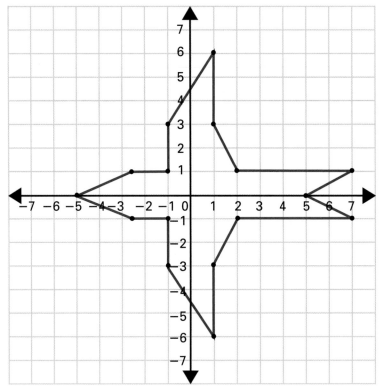

c. $(-2\frac{1}{2}, -1)$ $(-1, -1)$ $(-1, -3)$ $(1, -6)$ $(1, -3)$ $(2, -1)$ $(7, -1)$

Answers to Check Your Work

2. Compare your answer with a classmate. No, the statement is not always true. Sample response:

   The size of the figure will always change, except if you multiply all coordinates by 1 or −1.

3. You can get figure D by multiplying the coordinates of figure A by −1. If you did not find the answer to this question, make up an example for figure A. You could draw figure A as a rectangle and then draw figures B, C, and D following the instructions.

   Doing each multiplication separately has the same result as multiplying all the numbers and then making that change to the figure.

   $$2 \times -1 \times \tfrac{1}{2} = -1$$